Givanildo Sousa Gonçalves

Spodoptera frugiperda (J. E. SMITH) and Helicoverpa armigera (HÜBNER)

Givanildo Sousa Gonçalves

Spodoptera frugiperda (J. E. SMITH) and Helicoverpa armigera (HÜBNER)

Identification, monitoring and control strategies

ScienciaScripts

Imprint
Any brand names and product names mentioned in this book are subject to trademark, brand or patent protection and are trademarks or registered trademarks of their respective holders. The use of brand names, product names, common names, trade names, product descriptions etc. even without a particular marking in this work is in no way to be construed to mean that such names may be regarded as unrestricted in respect of trademark and brand protection legislation and could thus be used by anyone.

Cover image: www.ingimage.com

This book is a translation from the original published under ISBN 978-3-330-75882-7.

Publisher:
Sciencia Scripts
is a trademark of
Dodo Books Indian Ocean Ltd. and OmniScriptum S.R.L publishing group

120 High Road, East Finchley, London, N2 9ED, United Kingdom
Str. Armeneasca 28/1, office 1, Chisinau MD-2012, Republic of Moldova, Europe
Printed at: see last page
ISBN: 978-620-8-30280-1

SUMMARY

THANKS

Our thanks go to everyone who contributed directly or indirectly to the construction of this work.

INTRODUCTION

With grain production in the 2015/2016 harvest surpassing 200 million tons on 58.6 million hectares (CONAB, 2016), Brazilian agriculture is the second most emerging in the world, behind only the United States of America.

In Brazil, agriculture occupies large areas, and despite extensive cultivation, average productivity is still very low. Several factors are responsible for the low productivity per area of cultivation, with pests playing a major role. Among the main pests that attack agricultural crops, we can mention the caterpillars *Spodoptera frugiperda* (J. E. Smith) (Lepidoptera: Noctuidae) and *Helicoverpa armigera* (Hübner) (Lepidoptera: Noctuidae), as the pests that cause the greatest economic damage to producers. According to Gallo et al. (2002), *S. frugiperda*, by destroying the leaves, can reduce corn yields by up to 20%, with the critical period for its attack being the time close to flowering.

The caterpillars *of S. frugiperda* attack the stalk and can destroy it completely. They can also damage the ear, attacking and destroying it, preventing the grain from forming and reducing yields. The caterpillars *of H. armigera* feed on all parts of the plant, destroying the leaf area that is useful for photosynthesis, reducing productivity and causing high losses for producers.

Spodoptera frugiperda, also known as the corn earworm, is an insect that feeds on the main agricultural crops in America. Outbreaks occur regularly in corn-producing areas and populations commonly reach or exceed levels of economic damage (CRUZ, 2008). Among the main factors contributing to the increase and persistence of the pest in crops is the second "off-season" corn crop (Cruz, 1993), and intercropping.

Helicoverpa armigera (Hübner) (Lepidoptera: Noctuidae) is responsible for causing major economic losses in crops all over the world. This insect has a polyphagous feeding habit and feeds on more than 100 species of cultivated and non-cultivated plants, including forest species. It has a wide geographic distribution, with no record of its occurrence on the American continent until it was reported in Brazil in 2013 (CZEPAK et al., 2013).

The limited knowledge about the factors that influence the dispersal of this species, adaptive physiology, inheritance of resistance and efficient integrated measures for its control, currently carried out almost entirely with the intense use of insecticides, has led some populations to acquire resistance to molecules, especially those belonging to the pyrethroid chemical group (PATIL et. al., 2006).

The high migration capacity of *Helicoverpa armigera* adults, which can reach up to 1,000 km, has led to extensive reproductive crossbreeding, which has formed physiologically different populations that can be susceptible or resistant to standard control tactics. *Helicoverpa armigera* resistance to 640 insecticides has already been recorded worldwide (WYCKHUYS et. al., 2013). This is considered to be a serious problem that makes it difficult to control this pest, and prevention measures are needed to monitor adult populations and their dispersal.

The chemical control method is the most widely used against *S. frugiperda* and *H. armigera*, but often the expected results are not obtained. In addition, the occurrence of populations of both species resistant to some pesticides has already been recorded (DIEZ-RODRIGUEZ and OMOTO, 2001; WYCKHUYS et. al., 2013).

CHAPTER 1

1.1 *Spodoptera frugiperda* (J. E. SMITH) (LEPIDOPTERA: NOCTUIDAE)

Insects are considered pests when their population increases to the point of causing damage to plants, reducing their productivity.

Spodoptera frugiperda is native to the tropical and subtropical zones of the Americas, but can also be widely distributed throughout the temperate zones of the North American continent, during the spring and summer periods (SANTOS et al., 2004).

In Brazil, it stands out as one of the main pests of the corn crop and is commonly referred to as the corn earworm. The insect also attacks and causes damage to several other crops of economic importance, such as cotton, rice, alfalfa, peanuts, pumpkin, potatoes, cabbage, spinach, beans, cabbage, sorghum, wheat and tomatoes (EMBRAPA, 2004). *S. frugiperda* attacks have been recorded on more than 80 species of cultivated and non-cultivated plants.

In maize, the pest attacks at all stages of the crop's development, causing damage by destroying the leaves and especially the cartridge, reducing the photosynthesis area and consequently the yield. In bean and soybean crops, the caterpillars occur at any time the plant is cultivated and their attack can start as early as the first few days after plant emergence, when the plants are very sensitive to defoliation (EMBRAPA, 2003).

In 1999, *S. frugiperda* already caused losses of US$ 400 million worldwide. It is possible that these losses have exceeded US$ 1 billion annually since 2010. Its occurrence has been recorded in all Brazilian states, causing productivity losses of around 50%.

The high supply of food throughout the year for *S. frugiperda*, or as the host plants are known, is undoubtedly one of the main factors contributing to the persistence of the pest year after year in Brazilian crops. The high technology used in various crops in some regions of Brazil, such as Paranà, São Paulo, Rio Grande do Sul and parts of the Midwest, has made it possible to grow corn, cotton, rice, beans and other crops all year round through irrigation, increasing the source of food for the caterpillars to perpetuate themselves in the producing regions during the off-season, making it possible for caterpillar populations to rise rapidly in the following harvests.

The movement of *S. frugiperda* caterpillars between crops is another concern for producers, since the phenology of the crops is different, allowing them to migrate en masse between the different crops. This same package of *S. frugiperda* host plants also includes cover crops such as sorghum, millet, forage peanuts and others. This fact may also have contributed to an increase in the number of plants that are no longer sporadic food but become a permanent diet for *S. frugiperda*.

The high populations of caterpillars of this pest often destroy part of the pastures used by livestock farmers. This destructive ferocity has led producers to make uncontrolled use of insecticides to control populations, causing the pests to select or become resistant to different chemical molecules. The application of insecticides without a prior survey of the pest and the molecule's mechanism of action is an increasingly notorious problem in the environment, since non-selective molecules can kill many other insects besides the target pest, including the pests' natural enemies. The most efficient and correct method is integrated pest management (IPM), where a series of measures are adopted in order to reduce the different populations to levels that do not cause economic damage.

Among the measures used in the IPM *of S. frugiperda*, we highlight the use of insecticides that act on specific mechanisms of the target insects. In the case of caterpillars, a good option is chitin synthesis inhibitors, insecticides that only have an effect on lepidoptera, as they act on the insect's ecdysis. Another measure is crop rotation, with the use of non-host crops for the pest in the inter-harvest period. And currently, a measure that is widely used in IPM and that has allowed producers to feel comfortable is the use of genetically improved plants, which have in their DNA one or more genes from the bacterium *Bacillus thuringiensis* that express the production of an insecticidal toxin in the plant, causing the death of different species of lepidoptera.

1.2 **Identification of** *Spodoptera frugiperda*

Figure 1. male above and female below of *S. frugiperda*. Photo: Ivan Cruz

S. frugiperda moths can be found on the leaves of plants, close to the ground or between the leaves during the day. The main difference between the sexes in the moths is the color of the first pair of wings, which are straw-colored in the female and darker and mottled in the male (Figure 1). Two other basic features are the abdomen and the anal opening. In females, the abdomen is wider in the middle and the anal opening is a straight longitudinal line, while in males it is a straight longitudinal line. The moths can reach a wingspan of 3.5 cm and lay more than 2,000 eggs (CRUZ et al., 2008).

The color of the eggs can vary, but they are usually greyish. They are laid on the upper surface of the leaves, in layers ranging from one to four, and can hatch 3 to 4 days after oviposition. Once hatched, the 1st instar caterpillars head for the corn stalk, where they feed on the upper part of the youngest leaves of the plants, by scraping, which is a characteristic of the species, but after growth the caterpillar feeds on the whole plant (Figure 2).

Figure 2 - Corn plant attacked by *Spodoptera frugiperda*. Photo: H. Valicente.

After the first ecdysis, the caterpillar measures around 4 mm, has a whitish appearance, with a brownish back and an inverted "Y" on its head, which is the main way of identifying the species in its larval form. The larval period of *Spodoptera frugiperda* is between 12 and 30 days depending on the regional climate and feeding, where the caterpillars can reach a size of 50 mm (Figure 2). This period of the insect's life is divided into instars, which normally range from the 1st to the 7th, and can vary depending on the type of food and the climate. In an experiment carried out by Rosa et al. (2012) evaluating the longevity of *S. frugiperda* on different strains of corn, they obtained a caterpillar life cycle ranging from 10 to 21 days, and concluded that some strains affect the biology of the species by reducing its life span. Some researchers are betting on the use of products that act to stiffen the cell wall in order to reduce the herbivory of the caterpillars, a classic example being the use of silicon in plant cultivation.

Figure 3 - *Spodoptera frugiperda* caterpillar. Photo: Heraldo Negri.

The intensity of herbivory by *S. frugiperda* depends on the type of food and the larval period the species is in. In a study carried out by Grützmacher et al. (1999), in which they evaluated the herbivory of *S. frugiperda* according to the period of development or instar and the sex of the caterpillars, fed on irrigated rice leaves, they found that there was no difference in feeding according to the sex of the caterpillars, only their size, with a larva in the 7th instar feeding 65 times more than one in the 3rd instar.

1.3 *Spodoptera frugiperda* **breeding method**

The initial establishment of the *S. frugiperda* colony should be done with at least 400 individuals, taking into account mortality at the time of field collection. Rearing should be carried out in a controlled environment, with an average temperature of 25°C varying by 2°C, relative humidity of 80%, which can vary by 10%, and a 12-hour photophase.

The caterpillars collected in the field should be individualized, one per transparent plastic pot with a lid, containing the artificial diet. In this pot, the caterpillars will remain until they pupate and hatch into adults. The adults should be taken to cages, which can be made from PVC pipes with a diameter of 100 mm, lined with filter paper or napkins, in which the artificial diet should be placed (Table 1). Once the caterpillars have hatched, they should be separated into a maximum of 10 per 50 mL jar containing the artificial diet. After the 3rd instar, the caterpillars are individualized one per plastic pot, the pots must be of greater resistance and contain the artificial diet. In this pot, the caterpillars remain until they pupate and hatch into adults, which are then taken to the cages, where the cycle begins again.

Table 1. Components and quantities for the formulation of an artificial diet to feed 3,600 *Spodoptera frugiperda* caterpillars.

Components	Quantity
Beans	2000 g

Brewer's yeast	608,4 g
Wheat germ	950,4 g
Ascorbic acid	61,2 g
Methylparahydroxybenzoate sodium	37,8 g
Sorbic acid	19,8 g
Agar	250 g
Formaldehyde 40%	49, ml
Propionic propionic + acid phosphate + water Water	49.8 ml 16000 ml

Adapted from I. Cruz

The artificial diet is prepared as follows: cook the beans until they are soft, blend them in a blender for 3 to 4 minutes with formaldehyde, propionic acid + phosphoric acid + water and water. Then add the other ingredients with the exception of agar, which should be brought to the boil with water for 3 minutes and then added to the blender, blend everything for 8 minutes and put in the oven for 3 hours at 105°C, store in the fridge.

1.4 *Spodoptera frugiperda* **control methods**

Three main methods are used to control *Spodoptera frugiperda* populations: chemical control, biological control and the use of resistant or tolerant plants, including genetically modified ones.

1.4.1 **Chemical control**

Modern agriculture is marked by the cultivation of extensive areas, mainly with monocultures of strategic importance in animal and human nutrition. The chemical control of pests has become inevitable in the cultivation cycle, mainly due to the imbalance caused by pesticides in the relationships between insects and their natural enemies, providing conditions for the uncontrolled increase of populations and the inheritance of resistance acquired by pests over time of exposure to chemical products.

Several insecticides have been registered to control *S. frugiperda. frugiperda*, including Curyom 550 EC, Carbaryl Fersol 480 SC, Carbaryl Fersol Pó 75, Carbofuran Sanachem 350 TS, Endosulfan AG, Engeo Pleno, Diafuran 50,

Pirate, Atabron 50 CE, Chlorpyrifos Fersol 480 CE, Chlorpyrifos Sanachem 480 CE, Karate 50 CE, Klorpan 480 EC, Malathion 500 CE Sultox, Alsystin 480 SC and many other products. These products are usually applied in the form of aerial or ground sprays. The latter by means of self-propelled sprayers, trailed or mounted on the three points of the tractor.

1.4.2 **Biological control of** *S. frugiperda*

The biological control of *Spodoptera frugiperda* is carried out by the species' natural enemies such as predators, parasites and pathogens. Some basic requirements must be met for biological control agents to be used: specificity of the agent, compatibility, safety for humans and good control of the insect pest.

Among the natural enemies of *S. frufiperda* are baculoviruses, a group that includes the most studied viruses in pest control. This group of viruses has a single double strand of DNA capable of infecting a large number of insects. To date, 22 baculovirus strains have been isolated that are active in controlling *S. frugiperda*. The main form of application is via ground spraying or irrigation water where excellent control efficiency can be achieved. Valicente and Costa (1995) studied the efficiency of *Baculovirus spodoptera* applied via three irrigation water flushes of 3, 5 and 7 mm of water ha-1, at dosages of 3x101^2 , 3x10^{13} , 2.8x1012 polyhedra ha^{-1} of baculoviruses, and obtained an efficiency of over 77% control for all dosages and water flushes.

Another biological control agent for *S. frugiperda* of international importance are the *Trichogramma* wasps. These wasps parasitize the eggs of various orders of insects, and their efficiency is dependent on the type of host, the temperature and the fecundity of the female, but it is usually between 20 and 120 eggs per female. The greatest advance in the trade of wasps for the purpose of controlling pest insect populations has occurred in the last 20 years, mainly due to studies focused on laboratory breeding techniques based on artificial diets. Individuals are released at an average rate of 100,000 wasps per hectare, after which the application of non-selective insecticides should always be avoided.

There are several natural parasites and predators of *S. frugiperda*, which act to reduce the different populations of the pest in the field. Table 2 shows the parasites and Table 3 the predators, their respective orders and families.

Table 2. Main parasites *of Spodoptera frugiperda*.

Scientific name	**Family**	**Order**
Campoletis grioti	Ichneumonidae	Hymenoptera
Campoletis flavicincta	Ichneumonidae	Hymenoptera
Campoletis perdisstinctus	Ichneumonidae	Hymenoptera
Campoletis sonorrensis	Ichneumonidae	Hymenoptera
Archytas marmoratus	Tachinidae	Diptera
Lespesia archippivora	Tachinidae	Diptera
Eucelatoria sp.	Tachinidae	Diptera
Winthemia trinitatis	Tachinidae	Diptera
Archytas incertus	Tachinidae	Diptera
Euphorocera floridensis	Tachinidae	Diptera
Winthemia quadrispulata	Tachinidae	Diptera
Meteorus laphygmae	Braconidae	Hymenoptera
Meteorus autographae	Braconidae	Hymenoptera

Rogas laphygmae	Braconidae	Hymenoptera
Rogas terminaliss	Braconidae	Hymenoptera
Zelle mellea	Braconidae	Hymenoptera
Trichogramma fasciatum	Trichogramamatidae	Hymenoptera
Trichogramma minutum	Trichogramamatidae	Hymenoptera

Adapted from Cruz (2009).

Table 3. Main predators of *Spodoptera ssp.*

Scientific name	**Family**	**Order**
Coleomegilla maculata De G.	Coccinelidae	Coleoptera
Cycloneda sanguinea L.	Coccinelidae	Coleoptera
Eocanthecona sp.	Pentatomity	Hemiptera
Podisussagitta F.	Pentatomity	Hemiptera
Polistes versicolor Oliv.	Vespidity	Hymenoptera
Callosoma argulata (Chvr)	Carabidae	Coleoptera

Adapted from Cruz (2009).

CHAPTER 2

2.1 *Helicoverpa armigera* (HÜBNER) (LEPIDOPTERA: NOCTUIDAE)

Helicoverpa armigera (HÜBNER) (LEPIDOPTERA: NOCTUIDAE) has a wide geographical distribution, with its occurrence recorded in Europe, Asia, Africa and Oceania (ZALUCKI et al. 1986; GUO 1997). Until 2013, its attack had not been recorded on the American continent, so it was considered an A1 quarantine pest in Brazil, where little is known about this lepidopteran species.

According to Embrapa, the excessive population growth of caterpillars belonging to the *Helicoverpa* genus is the result of a cumulative process of inadequate cultivation and management practices. The main factor was undoubtedly the successive planting for many years of host species such as corn, soybeans, cotton and others, in large areas, with the adoption of improper management of the application of insecticides, making the environment favorable to the massive proliferation of insects, with food available, few natural enemies, their own places of reproduction and shelter during the off-season (EMBRAPA, 2013).

H. armigera is very polyphagous in its larval phase, with very rapid development, being able to complete its life cycle between 32 and 42 days, reproducing at extraordinary levels, reaching several generations in a year, its adults can fly more than 1,000 km, which gives it high dispersion in the different regions of the world. In addition, the molecular proximity between *Helicoperva* species, especially *H. zea* and *H. armigera*, can generate hybrid and fertile offspring, something that is of great concern to researchers and producers, as it is not known what the characteristics of this new individual will be, should this hypothesis be confirmed.

H. armigera caterpillars feed on leaves, fruit, inflorescences and stems. More than 100 species have been recorded as belonging to the pest's diet. Until

2007, *H. armigera* caterpillars caused US$ 5 billion in damage worldwide every agricultural year.

H. armigera caused damage close to R$1 billion in the 2013/2014 harvest alone in the state of Mato Grosso (IMEA, 2014). Its occurrence was recorded in the municipalities of Diamantina, Tangarà da Serra, Campo Verde, Rondonópolis and Alto Taquari, feeding on soybeans and cotton (SOSA-GÓMEZ et al., 2016). The cotton plant is sensitive to defoliation and can suffer losses from *Helicoverpa armigera* attacks of 10-15% in Bt plants (WideStrike®-WS) and 60-70% in conventional plants, requiring the application of high quantities of insecticides to control it (IMAMT, 2013), which is harmful to regional biodiversity.

Among the methods of controlling *Helicoverpa armigera* populations, chemical control is the most effective, as it is also the most damaging to the environment, followed by biological control using fungi and bacteria, the use of transgenic or non-transgenic resistant plants, and also the use of sex pheromones as a way of controlling this pest by confusing males and monitoring populations (EMBRAPA, 2013).

2.2 **Identification of** *H. armigera*

Identifying *H. armigera* is not so simple, as the caterpillars vary greatly in body color depending on the type of food, but in general they can show brown stripes on the sides of the thorax, abdomen and head (Figure 4). Its development in the caterpillar stage is divided into 7 distinct instars, which can last around 20 days, depending on climatic conditions and the type of food. After the fourth instar, *H. armigera* has dark abdominal tubercles that are clearly visible on the dorsal region. The morphological characteristics of this species can be identified in the field, but it must be correctly identified in the laboratory.

As adults, female *H. armigera* moths have yellowish forewings, while males have greenish-gray wings, darker in the distal third, and a small dark spot in the center of the wing. Females live an average of 11.7 days and males 9.2

days as adults. The sex pheromone used to attract the male is released 2 to 5 days after hatching, in the early hours of the morning. Copulation takes place during cooler, wetter periods and ceases during hotter, drier periods (KING, 1994).

Figure 4 - *H. armigera* caterpillar. Photo: Embrapa News Agency

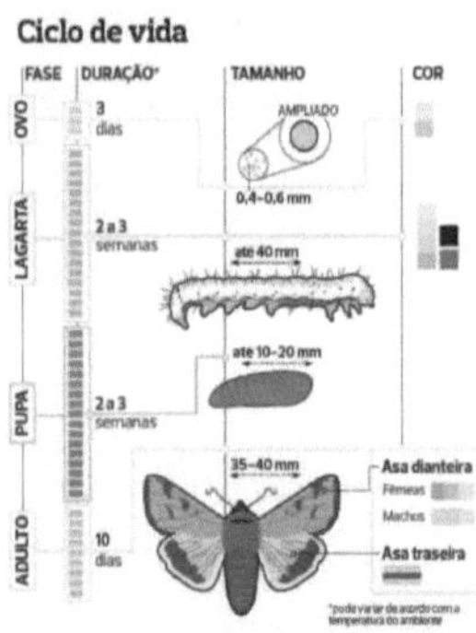

Figure 6. Life cycle of *H. armigera*. Illustration by Globo Rural.

The eggs of this species are bright yellowish-white in color, with shades of dark brown near the moment of hatching. The eggs are incubated for an average of 3.3 days, but can vary from 3 to 14 days, and vary in length from 0.4 mm to 0.6 mm and width from 0.4 mm to 0.5 mm (ALI and CHOUDHURY, 2009). Oviposition is usually carried out at night, placing them singly or in small groups, usually on the adaxial side of leaves, on stems, flowers and fruit.

2.3 *H. armigera* **breeding method**.

To establish an initial population of *H. armigera* for laboratory rearing, 300 to 500 caterpillars should be collected in the field and identified. The caterpillars will be separated one by one in a transparent plastic pot with a lid, containing the artificial diet, until they pupate and hatch into adults. Afterwards, the adults will be taken to the cages containing the artificial diet (Table 4). The cages can easily be made from PVC pipes with a diameter of more than 100 mm, lining their sides with napkins or filter paper, where the female will oviposit. After oviposition, remove the papers and transfer them to plastic jars with lids and wait for the caterpillars to hatch. The newly hatched caterpillars should then be separated into a maximum of 10 per transparent plastic pot with a lid. After reaching the 3rd instar, approximately one week later, the caterpillars should be individualized in 50 mL transparent jars with lids, containing the artificial diet, where they will remain until the adult hatches. After that, they will be taken back to the cages containing the artificial diet and the cycle begins again.

The breeding environment should be controlled with an average temperature of 27° C, relative humidity of 70% and a 12-hour photophase.

The artificial diet used to raise *H. armigera* is divided into three, two pasty for the caterpillar stage, the first until it reaches the 3rd instar (first week), the second from the 3rd instar to the pupa (Table 4 and 6), and another liquid for the adult stage (Table 5).

Table 4: Ingredients used to make the artificial diet for *H. armigera* caterpillars.

Ingredients in (g)	**Proportion 1st stage caterpillar**	**Proportion 2° caterpillar stage**
White beans	75,00	412,00
Wheat germ	60,00	329,60
Soybean meal	30,00	164,00
Milk powder	30,00	124,00
Brewer's yeast	37,50	206,00
Ascorbic Acid	3,60	19,60
Sorbic acid	1,80	10,00
Methylparahydroxybenzoate (Nipagin®)	3,00	23,20

Tetracycline	0,12	0,40
Agar	23,00	76,00
Ingredients Liquid (mL)		
Vitamin Solution	9,00	32,00
Formaldehyde	3,60	20,00
Fungus Inhibitor	3,00	24,00

Source: Embrapa (2014).

Table 5 Ingredients used to formulate the diet for adult *H. armigera*.

Ingredients	**Quantity**
Distilled water	1 liter
Ascorbic acid	1 gram
Sugar	25 grams

Source: Embrapa (2014).

Table 6. Ingredients used in the formulation of the dietary fungus inhibitor for *H. armigera*, caterpillar stage.

Ingredients	**Quantity in mL**
Propionic Acid	41,80
Phosphoric Acid	4,20
Water	54,00

Source: Embrapa (2014).

After purchasing the materials, follow the 10 steps described below: adapted from Embrapa (2014):

1- Cook the beans until soft, place on two sterilized aluminium trays in an oven at 75° C.

2- Prepare fungus inhibitor by adding water first, then acids, then place in glass container and store in fridge. Maximum shelf life 30 days.

3- Add water to the broth until it reaches 3 liters of liquid.

4- Cook the agar in 3 liters of liquid on the stove or in a microwave oven.

5- Put the beans in a blender and add 1 liter of water to blend.

6- Gradually add the solid ingredients and the cooked agar to the blender.

7- Add the formaldehyde and mold inhibitor last.

8- Remove the trays from the oven with suitable gloves and place them on a bench that can withstand high temperatures.

9- Distribute the diet on the trays (use a mask) and leave to cool. Once cool, place the trays in a plastic bag.

(10 kg bag), label them with the name of the diet and store in the fridge. Whenever you use the diet, remove the plastic, wipe off the excess water with a paper towel and leave it in the germicide for at least 25 minutes.

2.4 *H. armigera* control methods

The control measures used to reduce *H. armigira* populations are basically divided into three: chemical control, biological control and the use of genetically modified plants in association with integrated pest management (IPM).

2.5 Chemical control of *H. armigera*

Until 2013, there were no chemical products for controlling *H. armigera* in Brazil that were registered with the Ministry of Agriculture, Livestock and Supply (MAPA). With outbreaks occurring in different regions, causing huge losses for producers, MAPA issued more than 20 emergency regulations by the end of 2014. Regulations related to the declaration of a state of phytosanitary emergency, authorizing the import and application of chemical molecules (Table 3), registered in other countries such as the USA. However, the abusive application of insecticides should always be avoided, especially those that do not have specific mechanisms of action on the insect, causing the death of several other species, often even natural enemies of the lepidoptera complex. This undoubtedly causes environmental imbalance, increasing the risk of the species acquiring resistance or tolerance to different chemical molecules.

Table 3. Main chemical groups and active ingredients of insecticides used to control *H. armigera* populations.

Chemical group	Active ingredient
Spinosyns	Spinosad
Spinosyns	Spinetoran
Na blockers	Chlorfenapyr

Na blockers	Metaflumizone
Benzonylureas	Chlorfuazuron
Diaminas	Flubendiamide
Benzoylurea	Lufenuron
Diacylhydrazines	Methoxyfenozide
Benzonylureas + Carbamate	Novaluron + Methomyl
Benzonylureas + Organophosphate	Lufenuron + Profenofos
Carbamate	Methomyl

Products reported by the agrochemical industry to MAPA for emergency registration to control *Helicoverpa armigera* in soybean, cotton and corn crops.

2.6 **Biological control of** *H. armigera*

Biological control consists of reducing pest populations by using natural enemies as combatants, without the use of chemical products. In Brazil, there are currently not many natural control agents for *H. armigera* registered with MAPA. Another problem is the technology for mass production in laboratories, but there are already some viruses, parasites and predators that can be found for sale on the domestic market (Table 4).

Another method used to control *H. armigera* in corn, soybean and cotton crops is the use of resistant, genetically improved cultivars. Known worldwide as Bt technology, these are plants whose DNA contains genes from the bacterium *Bacillus thuringiensis*, which gives the plant the ability to produce toxins that are harmful to lepidopterans. Plants with this technology can produce two or more different types of insecticidal toxins (Table 5).

Table 4. Biological products for the management of *H. armigera*.

Product	**Species**
Microorganism: Bacteria	*Bacillus thuringiensis*
Microorganism: Entomopathogenic virus	*Baculovirus H. armigera* NPV
Microorganism: Virus entomopathogenic	*Baculovirus Helicoverpa zea* NPV
Beneficial insect - parasitoid	*Trichogramma pretiosum*

*Other products may be authorized or withdrawn from the market as of this date.

Table 5. Main Bt toxins present in the 2013/2014 crop in Brazil.

Culture	Toxins
Soy	CrylAc
Corn	Cry2Ab, Cry1Ab, Cry1A.105, Cry 1F, VIP3Aa
Cotton	Cry2Ab, Cry1Ab, Cry 2Ae, Cry1Ac, Cry1F

*Other products may be authorized or withdrawn from the market as of this date.

CHAPTER 3

3.1 *Spodoptera frugiperda* (J. E. SMITH) (LEPIDOPTERA: NOCTUIDAE), chemical control in corn, bean, soybean and sorghum.

3.1.1 Characterization of insecticides

Among the factors that determine the relative efficiency of insecticides in controlling caterpillars that cause damage to agricultural plants are the mechanisms of action which, according to Dadialla et al. (1998), can act more or less quickly on the mortality of the insect.

Busato et al. (2006) observed that insecticides that inhibit chitin synthesis act more slowly in terms of the disturbances caused to the insect's physiology, and neurotoxic insecticides act more quickly.

One day after spraying the insecticides on cotton, Vidal et al. (2005) found significantly fewer *Alabama argillacea cater*pillars in the plots containing the insecticide triazophos, with control efficiency of over 80%. According to the researchers, this result was to be expected, as this insecticide has a "shock" effect, while flubendiamide starts to reduce the caterpillar population from the third day after spraying.

The active insecticide ingredient flubendiamide, from the phthalic acid diamide chemical group, and chlorantraniliprole, from the anthranilamide chemical group, require some time to kill the insects, which means that their relative efficiency in killing them tends to occur after 72 hours. These insecticides activate ryanodine receptors by regulating calcium release in the sarcoplasmic reticulum of muscle cells, affecting insect muscle contraction (EBBINGHAUS-KINTSCHER et al., 2006; MASAKI et al., 2006; LÜMMEN et al., 2007; SATELLE, et al., 2008).

Other insecticides, such as diflubenzuron and methoxyfenozide, which belong to the Benzoylurea and Diacylhydrazine chemical groups, also generally require a longer period of time to effectively control the pest caterpillar population. According to Silva et al. (2003), the efficiency of diflubenzuron on the mortality of *Anticarsia gemmatalis* was seen a few days after it was sprayed on soybeans, because these chemical molecules have the characteristic of not having a "shock" effect, and of having a residual effect.

Benzoylurea insecticides inhibit the formation of chitin synthase from its zymogen by interfering with a protease responsible for activating chitin synthase (RETNAKARAN et al., 1985). The caterpillars cannot free themselves from their exocuticle because they are unable to secrete new endocuticle.

The insecticide methoxyfenozide, from the Diacylhydrazine group, is an ecdysis accelerator that mimics the natural insect hormone ecdysone. It acts by binding strongly to the ecdysone receptor protein, activating it and starting the ecdysis process. Immediately the caterpillars stop feeding and produce a new, but poorly formed, cuticle underneath the old one.

There are insecticides capable of acting more quickly on caterpillar mortality. This is the case with methomyl, the chemical group Oxime Methylcarbamate, which inhibits the enzyme acetylcholinesterase in nerve synapses, causing nerve impulses to pass continuously, leading the insect to fatigue and consequently death.

Alphacypermethrin and lambdacyhalothrin, a pyrethroid chemical group, also reduce the time it takes for insects to die. Acting on the sodium channels of the axon membrane, they reduce and delay the conductance of sodium into the cell, suppressing the efflux of potassium (GALLO et al., 2002). As a result, intermittent nerve impulses are generated.

The mixture of alfacipermethrin + teflubenzuron to control *Spodoptera frugiperda* at a dose of 16.5+16.5 g ha^{-1} of a.i. was efficient at three, seven and ten days after spraying. The use of alfacipermethrin (40 g.ha^{-1} of a.i.) gradually increased its efficiency, reaching just over 90% on the seventh day (BATISTA NETO et al., 2005).

The *Spodoptera frugiperda* caterpillar is a polyphagous agricultural pest, feeding on more than 80 plant species (CAPINERA, 2002; POGUE, 2002; BUSATO et al., 2002). This study is therefore important for screening groups of insecticides that will later be tested in the field. The aim was to evaluate the mortality of *S. frugiperda* fed on corn, bean, soybean and sorghum leaves immersed in solutions containing insecticides.

3.1.2 Maize cultivation

Corn is an annual crop belonging to the grass family, genus *Zea* and species *Zea plus*. Its probable center of origin is the American continent (CAMPOS et al., 1987).

It is a plant with a C4 metabolism, which gives this grass high mass production and a wide range of cultivation on the planet. It has separate male and female inflorescences, characterizing maize as a monoecious species. The kernel of this species is a fruit, called a caryopsis, covered by a thin colorless, red, brown or variegated layer derived from the ovary wall and called the pericarp (CASTRO et al., 1999).

Its greatest economic value lies in the kernels, which have a high energy and protein value, the fiber composition and the high dry matter production potential make it viable to use this species as a fodder plant for animal feed (CALONEGO et al., 2011).

However, in order to obtain good yields in different regions of Brazil, the cultivar's cycle and its appropriate phenological stages must be observed. According to Ritchie et al. (2003), the life cycle of the maize plant can be divided into the vegetative (V) and reproductive (R) phases, which are further subdivided into stages.

World corn production in the 2014/2015 harvest was 1 billion tons, with the US being the world's largest consumer, followed by China, the European Union and Brazil. Corn prices have varied widely, mainly due to stocks, productivity, imports, exports and consumption.

It is worth noting that production planning and market analysis are the main factors that define profitability or loss when growing corn.

"It is important to emphasize that producers must take into account the higher production costs, negotiating better prices for their production, which will allow them to pay these costs, and then negotiate the surplus at the best time" (CONAB, 2016).

3.1.3 Growing Beans

The common bean (*Phaseolus vulgaris L*) is one of the foods most consumed by Brazilians. In the 2013/2014 harvest in Brazil, production reached 3.4 million tons in an area of 3.3 million hectares. Compared to the 2012/2013 harvest, which produced 2.8 million tons in an area of 3.0 million hectares, national production grew by 13.2%.

The bean is a short-cycle species, 80 to 100 days depending on the cultivar, with a determinate or indeterminate growth habit, climbing or not, forming a canopy of 40 to 50 cm in height. It belongs to the legume family, genus *Phaseolus* and species *Phaseolus vulgaris L.* It probably originated in South America.

In recent years, Brazil has been the world's leading producer and consumer of common beans. Beans are grown in practically all Brazilian states, in the most varied climatic conditions and at different sowing times. In the central-western region, it is generally sown in three periods: the first "water" season; the second "dry" season and the third "winter" season, during which irrigation is necessary.

The phenological development of the bean plant is also divided into two stages, vegetative and reproductive (Table 6). This division is made in order to describe and characterize morphophysiological factors that occur at specific times in the crop's development.

Table 6. Phenological stages of bean development.

V0-Germination: absorption of water by the seed; emission of the radicle and caulicle and their transformation into a primary root.

V1-Emergence: the cotyledons appear at ground level, separate and the epicotyl begins to develop.

V2-Primary leaves: fully opened primary leaves.

V3-First trifoliate leaf: opening of the first trifoliate leaf and appearance of the second trifoliate leaf.

V4-Third trifoliate leaf: opening of the third trifoliate leaf and formation of branches in the buds of the lower nodes.

R5-Pre-flowering: appearance of the first flower bud and the first raceme. The flower buds of cultivars with determinate growth habit (type I) are formed at the last node of the stem and branch. In cultivars with indeterminate growth habit (Types II, III and IV) the racemes appear first at the lowest nodes.

A- Flowering: opening of the first flower

R7 - Pod formation: appearance of the first pod until it is 2.5 cm long, i.e. with a wilted corolla that is still attached or drooping.

R8 - Pod filling: the first pod begins to fill (seed growth). At the end of this stage, the seeds lose their green color and begin to show the characteristics of the variety. Defoliation begins.

R9-Maturation: the pods lose their pigmentation and begin to dry out. The seeds develop the typical color of the cultivar.

V = Vegetative; R = Reproductive; Each stage begins when 50% of the plants present the conditions relative to the stage; Source: Adapted from Fernàndez et al. (1985).

Various factors can influence the productivity and quality of the grain, such as the type of soil, the technology used by farmers, the climate and the sowing times. One of the most important is pest attacks, which can cause irreversible damage to the quality and productivity of the crop in a short space of time.

3.1.4 Soybean cultivation

The cultivation of soybeans (*Glycine max* (L) Merrill) in Brazil has been intensified since the 1990s, especially with advances in plant breeding programs and the large-scale introduction of no-till farming. Brazil is currently the world's second largest producer of the crop, with the USA in first place.

The world produced 312.362 million tons of soybeans in the 2015/2016 harvest, of which Brazil produced 95.63 million tons in an area of 33.17 million hectares, with an average yield of 2,882 kg per hectare. The top producing states in Brazil were Mato Grosso, Paranà and Rio Grande do Sul.

Exports of Brazilian soybeans in the form of grains in 2016 amounted to 54.3 million tons, and the total generated by the sum of exports of grains, bran and oil was U$ 28 billion. The economic and social importance of growing this crop is remarkable. In many cities in Mato Grosso, such as Sorriso and Primavera do Leste, the municipality's greatest economic strength is the result of planting and trading soybeans.

Soya is a short-day plant, but due to extensive studies in genetic improvement, it can now be grown in many parts of the world. Its development is divided into two phases, vegetative and reproductive. Each of these phases is subdivided, the vegetative represented by the V and the reproductive by the R.

Description adapted from RITCHIE et al. (1982).

Vegetative phase

VC. From emergence to open cotyledons.

V1. First node; open unifoliolate leaves.

V2. Second node; first trefoil open.

V3. Third node; second trefoil open.

Vn. The umpteenth (last) open trefoil node before flowering.

Reproductive stage (Observation on the main stem).

R1. Start of flowering: up to 50% of plants with flowers.

R2. Full bloom: most racemes with open flowers.

R3. End of flowering: flowers and pods up to 1.5 cm long.

R4. Most of the pods in the upper third are 2-4 cm.

R5.1 Grains perceptible to the touch at 10% of the grain.

R5.2 Most pods with 10%-25% grain.

R5.3 Most pods between 25% and 50% grain.

R5.4 Most pods between 50% and 75% grain.

R5.5 Most pods between 75% and 100% grain.

R6. 100% grain pods and green leaves.

R7.1. onset of 50% yellowing of leaves and pods.

R7.2 Between 51% and 75% yellow leaves and pods.

R7.3 More than 76% yellow leaves and pods.

R8.1 Start at 50% defoliation.

R8.2 More than 50% pre-harvest defoliation.

R9. Harvest maturity point.

3.1.5 Sorghum cultivation

Sorghum Vulgare Pers. is a plant of tropical origin, it belongs to the grasses and can grow up to 5 meters high. It is cultivated in many parts of the world, at altitudes of up to 1800 meters and temperatures ranging from 20 to 30°C. It is not very demanding in terms of soil fertility, growing well in sandy soils.

Sorghum is widely used as animal feed in the form of silage, mainly for dairy cattle. It is also used as a ground cover, due to its high forage production, and is widely used to grow soybeans, cotton, corn and beans in a no-till system.

Sorghum cultivation is fully mechanized and uses the same planting, cultivation and harvesting equipment as soybean, corn and bean crops. But it can also be grown manually by small farmers who don't have much equipment.

3.2 Experimental research methodology

The laboratory test was carried out in five replications using a randomized block design and consisted of four insecticides (Table 7) plus a "control" used as statistical treatments in solutions for soaking corn and bean leaf sections, = fed to *S. frugiperda* caterpillars in order to compare population mortality.

Table 7. Insecticides used in the soaking solutions for corn, bean, soybean and sorghum leaf sections.

Insecticides	Active ingredient/chemical group	Dose (L ha^{-1})*
Witness	-	-
Curyom 550 EC	Lufenurom (500 g i.a L^{-1}) / Benzoylurea + Profenofos (50gi . aL^{-1})/ Organophosphate	0350
Endosulfan AG	Endosulfan (635gi . aL^{-1})/ Cyclodienochlorate	1,500
Full Engeo	Lambda-Cyhalothrin (106 g i. aL^{-1})/ Pyrethroid + Thiamethoxam (146 g i.a L^{-1}) / Neonicotinoid	0,150
Klorpan 480 EC	Chlorpyrifos (489. 80gi. aL^{-1})/ Organophosphate	0,500

*Liters of commercial product used per hectare in 600 liters of water.

The experimental plots consisted of 19-centimeter Petri dishes lined with moistened filter paper. The filter paper was moistened with 1.0

mL of distilled water were placed on 10 1st instar caterpillars (experiment 1) and 3 3rd instar caterpillars (experiment 2) provided by the biolaboratory of the Centro Universitario da Grande Dourados (UNIGRAN). The experimental plots were kept on a marble bench in an air-conditioned room set at 25 ± 2° C, 12-hour photophase. Glass lids were used to close the plates. The leaf sections from the plant species fed to the caterpillars were changed every 24 hours, always previously immersed in solutions containing the insecticides described in Table 7.

The mortality of *S. frugiperda* caterpillars was assessed daily until the third day after the trial was set up. Relative efficiency was calculated according to (ABBOTT, 1925).

Relative efficiency is mandatory in the reports on Agronomic Efficiency and Practicability tests for pesticides and related products sent to the Ministry of Agriculture, Livestock and Supply (MAPA) for the registration of new products (SDA Normative Instruction No. 36 of November 24, 2009). This public body considers treatments with positive control values of more than 80% in relation to the control to be efficient.

Most of the time, therefore, relative efficiency explains better and/or is more valid than the suggestions of average tests, for example: the fact that a treatment differs or not from the control does not mean that it is efficient. Therefore, in this study, treatments with a relative efficiency on caterpillar mortality above 80% were considered efficient.

Statistical analyses were carried out using the Genes (CRUZ, 2001) statistical software application.

3.3 Results and Discussion

The relative efficiency of the insecticides tested on the mortality of agricultural pests is related to their modes of action and/or absorption by the insect, which may or may not have an immediate toxic action with an influence on its development, residual, associations in the mixture and the type of formulation of these products (CESSA et al., 2013). In addition, according to these authors, the amount of the insecticide's active ingredient deposited on the leaf is also a determining factor in insect toxicity. Therefore, in this study, as the method of depositing the insecticides on the leaves does not represent those established by the manufacturers of these products for insect control, it was agreed that this factor was not relevant to the discussion of the results.

In general, up to 48 hours after the trial was set up, 100% relative efficiency was observed in the mortality of 1st and 3rd instar *S. frugiperda* caterpillars when fed corn and soybean leaves immersed in a solution containing the insecticides evaluated (Tables 8, 9, 10 and 11).

When fed bean and sorghum leaves immersed in a solution containing the insecticide Engeo Pleno, there was 100% relative efficiency in the mortality of 1st instar *S. frugiperda* caterpillars up to 72 hours after the trial was set up. This same behavior was observed for sorghum using the insecticide Endosulfan AG. The other insecticides promoted the mortality of 1st instar caterpillars up to 48 hours after the trial was set up. For 3rd instar *S. frugiperda* caterpillars, all the insecticides tested had a relative efficiency of 100% on the population of these caterpillars up to 48 hours after the trial was set up (Tables 12, 13, 14 and 15).

Table 8. Relative mortality efficiency of *S. frugiperda* caterpillars fed on corn leaf sections immersed in insecticides 24, 48 and 72 hours after the experiment was set up, calculated on the basis of caterpillar mortality - 1st instar caterpillars.

Insecticides	Active ingredient/chemical group	24 h	48 h	72 h
Efficiency				I (%)
Curyom 550 EC	Lufenurom (500 g i.a L^{-1}) / Benzoylurea + Profenofos (50gi . aL^{-1})/ Organophosphate	100	-	-
Endosulfan AG	Endosulfan (635gi . aL^{-1})/ Cyclodienochlorate	100	-	-
Full Engeo	Lambda-Cyhalothrin (106 g i. aL^{-1})/ Pyrethroid + Thiamethoxam (146 g i.a L^{-1}) / Neonicotinoid	60	100	-
Klorpan 480 EC	Chlorpyrifos (489. 80gi . aL^{-1})/ Organophosphate	100	-	-

*Dose of commercial product used per hectare in 600 l of water Source: Own elaboration

Table 9. Relative mortality efficiency of *S. frugiperda* caterpillars fed on corn leaf sections immersed in insecticides 24, 48, and 72 hours after the experiment was set up, calculated on the basis of caterpillar mortality - 3rd instar caterpillars.

Insecticides	**Active ingredient/chemical group**	**24 h**	**48 h**	**72 h**
		Efic	**ience (%)**	
Curyom 550 EC	Lufenurom (500 g i.a L^{-1}) / Benzoylurea + Profenofos (50gi . aL^{-1})/ Organophosphate	67	100	-
Endosulfan AG	Endosulfan (635gi . aL^{-1})/ Cyclodienochlorate	100	-	-
Full Engeo	Lambda-Cyhalothrin (106 g i. aL^{-1})/ Pyrethroid + Thiamethoxam (146 g i.a L^{-1}) / Neonicotinoid	33	100	-
Klorpan 480 EC	Chlorpyrifos (489. 80gi. aL^{-1})/ Organophosphate	100	-	-

*Dose of commercial product used per hectare in 600 l of water Source: Own elaboration

Table 10. Relative mortality efficiency of *S. frugiperda* caterpillars fed soybean leaf sections dipped in insecticides 24, 48, and 72 hours after the experiment was set up, calculated on the basis of caterpillar mortality - 1st instar caterpillars.

Insecticides	**Active ingredient/chemical group**	**24 h**	**48 h**	**72 h**
		Efficiency (%)		
Curyom 550 EC	Lufenurom (500 g i.a L^{-1}) / Benzoilurea + Profenofos (50gi. aL^{-1})/ Organophosphate	100	-	-
Endosulfan AG	Endosulfan (635gi . aL^{-1})/ Cyclodienochlorate	80	100	-
Full Engeo	Lambda-Cyhalothrin (106 g i. aL^{-1})/ Pyrethroid + Thiamethoxam (146 g i.a L^{-1}) / Neonicotinoid	100	-	-
Klorpan 480 EC	Chlorpyrifos (489. 80gi. aL^{-1})/ Organophosphate	100	-	-

*Dose of commercial product used per hectare in 600 l of water Source: Own elaboration

Table 11. Relative mortality efficiency of *S. frugiperda* caterpillars fed on soybean leaf sections dipped in insecticides 24, 48, and 72 hours after the experiment was set up, calculated on the basis of caterpillar mortality - 3rd instar caterpillars.

Insecticides	Active ingredient/chemical group	24 h	48 h	72 h
		Efficiency (%)		
Curyom 550 EC	Lufenurom (500 g i.a L^{-1}) / Benzoylurea + Profenofos (50gi .aL^{-1}) / Organophosphate	67	100	-
Endosulfan AG	Endosulfan (635gi . aL^{-1})/ Cyclodienochlorate	100	-	-
Full Engeo	Lambda-Cyhalothrin (106 g i. aL^{-1})/ Pyrethroid + Thiamethoxam (146 g i.a L^{-1}) / Neonicotinoid	33	100	-
Klorpan 480 EC	Chlorpyrifos (489. 80gi. aL^{-1})/ Organophosphate	100	-	-

*Dose of commercial product used per hectare in 600 l of water Source: Own elaboration

Table 12. Relative mortality efficiency of *S. frugiperda* caterpillars fed bean leaf sections immersed in insecticides 24, 48 and 72 hours after the experiment was set up, calculated on the basis of caterpillar mortality - 1st instar caterpillars.

Insecticides	Active ingredient/chemical group	24 h	48 h	72 h
		Efficiency (%)		
Curyom 550 EC	Lufenurom (500 g i.a L^{-1}) / Benzoylurea + Profenofos (50 g i.a L^{-1}) / Organophosphate	100	-	-
Endosulfan AG	Endosulfan (635gi . aL^{-1})/ Cyclodienochlorate	100	-	-
Full Engeo	Lambda-Cyhalothrin (106 g i. aL^{-1})/ Pyrethroid + Thiamethoxam (146 g i.a L^{-1}) / Neonicotinoid	60	90	100
Klorpan 480 EC	Chlorpyrifos (489. 80gi . aL^{-1})/ Organophosphate	100	-	-

*Dose of commercial product used per hectare in 600 l of water Source: Own elaboration

Table 13. Relative mortality efficiency of *S. frugiperda* caterpillars fed bean leaf sections dipped in insecticides 24,48, and 72 hours after the experiment was set up, calculated on the basis of caterpillar mortality - 3rd instar caterpillars.

Insecticides	Active ingredient/chemical group	24 h	48 h	72 h
		Efficiency (%)		
Curyom 550 EC	Lufenurom (500 g i.a L^{-1}) / Benzoylurea + Profenofos (50 g i.a L^{-1}) / Organophosphate	100	-	-
Endosulfan AG	Endosulfan (635gi . aL^{-1})/ Cyclodienochlorate	67	100	-
Full Engeo	Lambda-Cyhalothrin (106 g i. aL^{-1})/ Pyrethroid + Thiamethoxam (146 g i.a L^{-1}) / Neonicotinoid	67	100	-
Klorpan 480 EC	Chlorpyrifos (489. 80gi . aL^{-1})/ Organophosphate	67	100	-

*Dose of commercial product used per hectare in 600 l of water Source: Own elaboration

Table 14. Relative mortality efficiency of *S. frugiperda* caterpillars fed on sorghum leaf sections immersed in insecticides 24, 48 and 72 hours after the experiment was set up, calculated on the basis of caterpillar mortality - 1st instar caterpillars.

Insecticides	Active ingredient/chemical group	24 h	48 h	72 h
		Efficiency (%)		
Curyom 550 EC	Lufenurom (500 g i.a L- 1) / Benzoilurea + Profenofos (50 gi. aL^{-1}) / Organophosphate	100	-	-
Endosulfan AG	Endosulfan (635gi . aL^{-1})/ Cyclodienochlorate	90	-	100
Full Engeo	Lambda-Cyhalothrin (106 g i. aL^{-1})/ Pyrethroid + Thiamethoxam (146 g i.a L^{-1}) / Neonicotinoid	80	90	100
Klorpan 480 EC	Chlorpyrifos (489. 80gi . aL^{-1})/ Organophosphate	100	-	-

*Dose of commercial product used per hectare in 600 l of water Source: Own elaboration

Table 15. Relative mortality efficiency of *S. frugiperda* caterpillars fed on sorghum leaf sections immersed in insecticides 24, 48 and 72 hours after the experiment was set up, calculated on the basis of caterpillar mortality - 3rd instar caterpillars.

Insecticides	Active ingredient/chemical group	24 h	48 h	72 h
		Efficiency (%)		
Curyom 550	Lufenurom (500 g i.a L^{-1}) / Benzoylurea +	67	100	-

EC	Profenofos (50 g i.a L^{-1}) / Organophosphate			
Endosulfan AG	Endosulfan (635gi . aL^{-1})/ Cyclodienochlorate	100	-	-
Full Engeo	Lambda-Cyhalothrin (106 g i. aL^{-1})/ Pyrethroid + Thiamethoxam (146 g i.a L^{-1}) / Neonicotinoid	33	100	-
Klorpan 480 EC	Chlorpyrifos (489. 80gi . aL^{-1})/ Organophosphate	100	-	-

*Dose of commercial product used per hectare in 600 l of water Source: Own elaboration

The similar behavior observed with regard to the time of relative efficiency in the mortality of the population of 1st and 3rd instar *Sfrugiperda* caterpillars on different plant species is possibly related to the fact that it is a polyphagous pest and that the age of the caterpillar, or rather its instars, according to Silva et al. (2012), does not interfere with its food preference.

According to Dadialla et al. (1998), the difference between the products tested in terms of their relative effectiveness in killing *S. frugiperda* during the periods evaluated is due to their different mechanisms of action.

In this study, all the insecticides tested have a neurotoxic mechanism of action which, according to Busato et al. (2006), acts more quickly on the death of the insects, while the chitin synthesis inhibitors act more slowly in terms of the disturbances caused to their physiology. This may have contributed to the relative rapidity with which 100% of the 1st and 3rd year *S. frugiperda* caterpillars were killed.

The speed at which insecticides act more or less slowly on the mortality of insect pest populations has been observed in various studies. Cessa et al. (2013) found that neurotoxic insecticides with a "shock" action, when used alone or in a mixture with other insecticides, generally had a relative efficiency of up to 48 hours on the mortality of *S. frugiperda*, and that insect growth regulators used alone had a relative efficiency of over 80% on the mortality of *S. frugiperda* caterpillars from the third day onwards.

In the studies by Grutzmacher et al. (2000) and Tomquelski and Martins

(2007), the efficiency of insecticides belonging to the benzoylurea group was not found for controlling *S. frugiperda* three days after spraying, but after the seventh day their efficiency is over 80%.

Diflubenzuron, a physiological insecticide, was relatively effective in controlling the *S. frugiperda* population in the evaluation 15 days after spraying in the field. Viana and Costa (1998) attribute this to the residual "power" of this insecticide, due to its low solubility in water and its inhibiting action on the formation of new cuticle on the caterpillar, requiring more time to act on the insect's metabolism until it kills it.

In the work by Neto e Silva et al. (2001), it was found that insecticide growth regulators require more time to act on insects when compared to conventional products.

The efficiency of the chitin synthesis inhibitor active ingredients lufenurom (Match 50 EC) and novalurom (Rimon 100 EC) and the ecdysis accelerator active ingredients methoxyfenozide (Intrepid 240 SC) and tebufenozide (Mimic 240 SC) tested on *Grapholita molesta* caterpillars occurred in the evaluations carried out 14 days after their contact with the pest.

Arioli et al. (2004) found that methoxyfenozide was more than 80% effective at controlling *G. molesta* on peach fruit in the laboratory, after 96 hours. On the same pest and crop, Grützmacher et al. (1999) found better efficiency of this insecticide 14 days after its application.

In the work by Rodrigues et al. (2009), relative efficiency of over 80% was observed in the control of the *Alabama argillacea* population with the use of methomyl (172 g of i.a ha^{-1}) and *Bacillus thuringiensis* + methomyl (172 g of i.a ha^{-1}) two days after spraying, highlighting the "shock" effect provided by methomyl on the pest. For this evaluation period, the authors did not observe any relative efficiency with the use of the insecticide diflubenzurom.

3.4 Conclusions

In general, up to 72 and 48 hours after the trial was set up, 100% relative efficiency was observed in the mortality of 1st and 3rd instar *S. frugiperda* caterpillars, respectively, when fed on corn, soybean and bean leaves.

Engeo Pleno and Endosulfan AG were used for the longest period (72 hours after trial implementation) to achieve 100% relative efficiency in the mortality of 1st instar *S. frugiperda* caterpillars in bean and sorghum.

Among the factors relevant to the speed at which 100% relative efficiency is obtained in the mortality of 1st and 3rd *S. frugiperda* caterpillars, the mechanism of action can be cited as one of the most important.

CHAPTER 4

4.1 PRE-PROJECT GUIDE FOR A DOCTORATE IN AGRICULTURAL ENTOMOLOGY

4.1.1 Sex Pheromone: Management Strategy for *Helicoverpa armigera* (HÜBNER) (LEPIDOPTERA: NOCTUIDAE) in the cotton crop

4.1.2 Sexual pheromones

Pheromones are odors or chemical signals used in communication between insects in the exercise of their ecological relationships. Based on their behavioral response, these substances can be classified as sexual, aggregation, dispersal, alarm, territoriality and laying (VILELA and DELLA LUCIA, 2001). Sexual pheromones are those normally released by the opposite sex and their purpose is to attract the sexes for mating. The use of pheromones as a systematic approach to integrated pest management has become an excellent tool for monitoring the presence and density of insects in crops, providing knowledge for making the correct decision on when to control them, reducing unnecessary applications and environmental contamination (ZARBIN et al., 2009).

The use of pheromones in the monitoring and control of insects is a viable alternative for reducing environmental impacts, since this method is based on the monitoring and ecological control of insects through the confusion and non-recognition of females by males. A natural product based on sex pheromones, which can be obtained from production sites, is a viable option for reducing the use of synthetic chemicals and environmental pollution. This is an ecological strategy that can provide the productive sectors with alternatives that are less harmful to the environment and at the same time reduce expenditure on the purchase and application of synthetic chemicals that are aggressive to nature.

Monitoring *Helicoverpa armigera* and understanding its behavior are of great importance in understanding the factors involved in its distribution, predicting future infestations and determining the least aggressive control method for the environment. The aim of this proposal is to evaluate the response of *Helicoverpa armigera* to synthetic and natural sex pheromones and the optimum density of traps, as a strategy for the ecological management of the insect on cotton crops.

Table 16. Main chemical groups of pheromones used to monitor and control *H. armigera*.

Chemical group of the pheromone	**Active ingredient**
Unsaturated aldehydes	Z11-hexadecenal
Unsaturated aldehydes	Z9-hexadecenal
Semiochemical Iscalure armigera	(Z)-9-hexadecenal (Z90C16Ald;(Z)- 9-tetradecenal (Z9-C14Ald); (Z)-11-C16Ald)
Semiochemical-SPLAT-armigera	(Z)-9-hexadecenal (Z90C16Ald;(Z)- 9-tetradecenal (Z9-C14Ald); (Z)-11-C16Ald)
Semiochemicals + concentrated bait-Splat Noctutrap	(Z)-9-hexadecenal (Z90C16Ald;(Z)- 9-tetradecenal (Z9-C14Ald); (Z)-11-C16Ald)

*Products reported by the agrochemical industry to MAPA for emergency registration to control *Helicoverpa armigera* in soybean, cotton and corn crops.

4.2 Research objectives

To test, in laboratory and field trials, the response of *Helicoverpa armigera* to the commercial sex pheromone Iscalure Armigera and to pheromone compounds extracted from insect gandules.

Estimating the optimum number and spatial distribution of traps for the control of *Helicoverpa armigera* using sex pheromone in cotton crops.

4.3 Experimental research material and methods

This research proposal will be conducted in two experiments. Experiment 1 will be carried out in a laboratory. Experiment 2 will be conducted in an area cultivated with cotton on a farm located in the cerrado of Mato Grosso.

Experiment 1: In this part of the research, the objective is to evaluate the behavior of different ages of *Helicoverpa armigera* male adults in the perception of the commercial sex pheromone Iscalure Armigera and the pheromone compound extracted from the insects' glands. Iscalure Armigera is composed of the active ingredient (Z)-9-Hexadecenal (Z9-16; Ald); (Z)-11-Hexadecenal (Z11-16: Ald) 2.82 g kg $.^{-1}$

To establish the *Helicoverpa armigera* colony in the laboratory, caterpillars will be collected in the field from cotton plantations, following the specific characteristics of the insect described in technical circular 23 (EMBRAPA, 2013). The method of rearing and feeding *Helicoverpa armigera* in the juvenile and adult stages will follow the recommendations of technical circular 203 (EMBRAPA, 2014).

The behavioral bioassays of *Helicoverpa armigera* adults will be observed in a double-choice glass Y-type olfactometer, with a diameter of 4 cm, an initial arena of 22 cm, bifurcated into two arms of 24 cm each, using a method adapted from (GOKÇE et al., 2005). The tests will be conducted in a chapel with an air exhaust system, kept in an air-conditioned room. The stimuli used will be the synthetic sex pheromone Iscalure Armigera and pheromone compound extracted from the insects' glands, applied to (1.5 x 15 cm) paper. Adult males of *Helicoverpa armigera* aged 1, 5 and 10 days will be subjected to the synthetic and natural pheromone.

Responses will be considered positive when the insects have traveled at least 4 cm inside the arms containing the test treatment and remain in that area for at least 1 minute. A negative response will be considered if the insects walk the same distance and stay the same length of time in the arm containing the control treatment. No response will be recorded when the insects remain only in the main arm of the olfactometer. For each age of adult male *Helicoverpa armigera*, 20 repetitions will be carried out, with each repetition being observed for 20 minutes.

Experiment 2: The aim of this experiment will be to estimate the optimum number and spatial distribution in the field of Delta-type traps for monitoring *Helicoverpa armigera* in cotton plantations, attracted by the sex pheromone Iscalure armigera and pheromone compounds extracted from insect glands, correlating the population of males captured in each pheromone with the density of larvae in the cotton plants during a production cycle.

The design will be entirely randomized, in a 2x8 factorial scheme with two pheromones (synthetic and natural) and eight spatial distributions with five replications per treatment, the spatial distributions being 1 trap for 5 ha^{-1} , 1 trap for 4 ha^{-1} , 1 trap for 3 ha^{-1} , 1 trap for 2 ha^{-1} , 1 trap for 1 ha^{-1} , 2 traps for 1 ha^{-1} , 3 traps for 1 ha^{-1} , 4 traps for 1 ha^{-1} , the average spacing between traps was determined according to the recommendation for the pheromone Iscalure Armigera (MAPA, 001013 E). All the traps will be previously georeferenced to map the experimental area. The traps will be suspended from wooden stakes using a wire, at a height of approximately 1.50 m above the ground. The stakes will have an inverted L shape, with the longest part being 2.10 m and the shortest 0.5 m.

The pheromone lures will be replaced by a new one every 30 days and the traps and the density of larvae per 10 plants will be checked once a week. The data will be grouped into 30-day periods. To better interpret the results, climatic data will be collected on rainfall, relative humidity and temperature. The correlation between the number of adult insects sampled in the traps and the population of larvae in the cotton crop will be calculated using the simple correlation formula also used by (SHARMA et al., 2002).

The data will be submitted to analysis of variance and the means will be compared using the Tukey test at 5% probability, using the SISVAR computer resource (FERREIRA, 2011).

4.4 Expected results

The aim of this work is to learn and understand the behavior and response of

Helicoverpa armigera to synthetic and natural sex pheromones, the spatial distribution of caterpillars of the species and their relationship with the density of adults per area. We believe it is possible to determine the optimum number of traps that can provide the greatest sample representativeness without disproportionately increasing the cost of purchasing and maintaining the traps in the field.

We hope to monitor *Helicoverpa armigera* populations more efficiently, estimating possible future infestations based on the density and dispersal of adults captured. We hope to identify a natural product based on sex pheromone that can be obtained from production sites and that is highly efficient, reducing the use of synthetic chemicals, environmental contamination and unnecessary costs.

It is hoped that it will be possible to provide a technical and scientific basis for possible research aimed at using natural pheromones to monitor and control *Helicoverpa armigera* in different plant species. Strengthening the technique of monitoring and managing insects using pheromones, encouraging integrated management and raising awareness of the importance of developing a more sustainable agriculture.

The results obtained from the project are expected to generate a technological impact, due to the improvements obtained in insect management, as well as its practical functionality, which can guarantee advantages in the use of the method, social and political acceptability of the ecological practice studied, by improving the quality of life, reducing environmental contamination by pesticides, in addition to ecological preservation and balance.

BIBLIOGRAPHICAL REFERENCES

ABBOTT, W. S. A method of computing the effectiveness of on insecticide. **Journal of Economic Entomology**, Lanham, v.18, n.1, p.265-267. 1925.

ARIOLI, C. J.; BOTTON, M.; CARVALHO, G. A. Chemical control of *Grapholita molesta* (Busck) (Lepidoptera:Tortricidae) in peach trees. **Ciência Rural,** Santa Maria, v.34, n.6, p.1695-1700, 2004.

ALI, A.; CHOUDHURY, R.A. Some biological characteristics of *Helicoverpa armigera* on chickpea. **Tunisian Journal of Plant Protection**, v.4, n.1, p.99106, 2009.

BATISTA NETO, O. A.; PAULA, J. M. De.; PEIXOTO, M. F.; BARROS, E. M.; JAKOBY, G. L. Effect of different doses and insecticides on the control of *Spodoptera frugiperda* (J.E.Smith, 1797) (LEPDOPTERA: NOCTUIDAE) in the cotton crop. **In:** CONGRESSO BRASILEIRO DO ALGODÂO, 5., 2005, Salvador. Proceedings... Campina Grande: Embrapa Cotton, 2005. CD-ROM 078

BUSATO, G. R.; GRÜTZMACHER, A. D.; GARCIA, M. S.; ZOTTI, M. J.; NORNBERG, S. D.; MAGALHÂES, T. R.; MAGALHÂES, J. De BANDEIRAS. Susceptibility of caterpillars of the maize and rice biotypes of *Spodoptera frugiperda* (J.E. Smith, 1797) (Lepidoptera: Noctuidae) to insecticides with different modes of action. **Ciência Rural**, v.36, n.1 p.15-20, 2006.

CAMPOS, A. T.; FILHO, V. C. **Principais culturas II**. Câmara brasileira de livros. p.195-233. Campinas SP, 1987.

CALONEGO, J.C.; POLETO, L.C.; DOMINGUES, F.N.; TIRITAN, C.S. Corn productivity and growth in different plant arrangements. **Revista Agraria.** v.4, n.12, p.84-90. Dourados, 2011.

CAPINERA, J. L. **Handbook of vegetable pests**. San Diego: Academic Press, 2002.

CASTRO, P. R. C.; KLUGE, R. A. **Ecophysiology of annual crops**: Wheat,

Maize, Soy, Rice and Cassava. São Paulo. Nobel, 1999.

CESSA, R. M. A.; MELO, E. P. De.; LIMA JUNIOR, I. Dos S. De. Mortality of *Spodoptera frugiperda* (J. E. Smith) (Lepdoptera: noctuidae) fed on corn and bean leaves immersed in solutions containing insecticides. **Revista Agrogeoambiental**, v.5, n.1, p.1-7, 2013.

CHOUGULE, N. P.; GIRI, A. P.; SAINANI, M. N.; GUPTA, V. S. Gene expression patterns of *Helicoverpa armigera* gut proteases. **Insect Biochemistry and Molecular Biology**, v.35, p.355-367, 2005.

CONAB. Monitoring the Brazilian grain harvest, v. 1 - Crop 2013/14, n. 12 - Twelfth Survey, Sept. 2014.

CONAB. **Brazilian crop monitoring: grains**. 2015/2016 harvest, fourth survey 2016. Brasilia, DF, 2016. 154 p. Available at: http://www.conab.gov.br/OlalaCMS/uploads/arquivos/16 0112 09 00 46 boletim graos january 2016.pdf. Accessed on: 10/11/2016.

CRUZ, I. Management of corn crop pests. **In:** GALVAO, J.C.C.; MIRANDA, G.V. (Ed.). Corn production technologies: economics, cultivars, biotechnology, 2ª crop, fertilization, chemigation, diseases, weeds and pests. Viçosa, MG: Editora UFV, 2008, p.311-366.

CRUZ, I. **Recomendações técnicas para o cultivo do milho**: Principais pragas e controle. Brasilis: EMBRAPA-SPI, Document, 204p, 1993.

CRUZ, I. Rearing methods for entomophagous agents of *Spodoptera frugiperda. In*: BUENO, V.H.P. (ed). Biological Pest Control: Mass Production and Quality Control. Editora UFLA, Lavras, p. 237-275, 2009.

CZEPAK, C.; ALBERNAZ, K. C.; VIVAN, L. M.; GUIMARAES, H. O.; CARVALHAIS,T. First record of *Helicoverpa armigera* (Hübner) (Lepidoptera: Noctuidae) in Brazil. **Pesquisa Agropecuâria Tropical**, Goiânia, v. 43, n. 1, p. 110-113, 2013.

DADIALLA, T. S.; CARLSON, G. R.; LE, DAT P. New insecticides with ecdysteroidal and juvenile hormone activity. **Annual Review of Entomology**, v.43,n.1, p.545-569, 1998.

DIEZ-RODRIGUEZ, G.I.; OMOTO, C. Inheritance of resistance of *Spodoptera frugiperda* (J.E. Smith) (Lepidoptera: Noctuidae) to lambda-cyhalothrin. **Neotropical Entomology**, v.30, p.311-316, 2001.

EBBINGHAUS-KINTSCHER, U.; LUEMMEN, P.; LOBITZ, N.; SCHULTE, T.; FUNKE, C.; FISCHER, R.; MASAKI, T.; YASOKAWA, N.; TOHNISHI, M. Phthalic acid diamides activate ryanodine sensitive Ca^{2+} release channels in insects. **Cell Calcium,** v.39, n.1, p.21-33, 2006.

EMBRAPA. **Biological control of the corn rootworm** *Spodoptera frugiperda* **using the egg parasitoid** Trichogramma pretiosum. Technical communication 114. Sete Lagoas MG. 2004.

EMBRAPA. Production systems 2. Cultivation of cowpeas. Electronic version. 2003. Available at:

http://sistemasdeproducao.cnptia.embrapa.br/FontesHTML/Feijao/FeijaoCaupi/. Accessed on 20/12/2014.

FANCELLI, A. L.; NETO, D.D. **Feijao Production**. Piracicaba SP. Livroceres, p 33-45, 2007.

FERNANDEZ, F.; GEPTS, P.; LÓPEZ, M. Stages of development in the frijol plant. In: LÓPEZ, M.; FERNANDEZ, F.; SCHIINHOVEN, A. van. **Frijol: Research and production**. Santiago de Chile, p.61-78, 1985.

FERREIRA, D.F. Sisvar: a computer system for statistical analysis. **Ciências Agrotécnicas**, Lavras, v.35, n.6, p.1039-1042, 2011.

FIGUEIREDO, M.L.C.; PENTEADO-DIAS, A. M.; CRUZ, I. **Damage caused by** *Spodoptera* **to dry matter production and grain yields in maize** (Technical Communication, 130). Embrapa, Sete Lagoas, MG: CNPMS, 2005. 6p.

GALLO, D.; NAKANO, O.; SILVEIRA NETO, S.; CARVALHO, R. P. L.; BAPTISTA, G. C. de; BERTI FILHO, E.; PARRA, J. R. P.; ZUCCHI, R. A.; ALVES, S. B.; VENDRAMIM, J. D.; MARCHINI, L. C.; LOPES, J. R. S.; OMOTO, C. **Entomologia agricola**. Piracicaba: Fealq, 2002. 920 p.

GOKÇE, A.; STELINSKI, L.L.; WHALON, M.E. Behavioural and electrophysiological responses of leafroller moths to selected plant extracts. **Environmental Entomology,** Oxford, v.34, p.1426-1432, 2005.

GRÜTZMACHER, A. D.; MARTINS, J. F. S.; AZEVEDO, R.; GIOLO, F. P. Effect of insecticides and application technologies on the control of the cartridge caterpillar in the corn crop in the vàrzea agroecosystem. **In:** ANNUAL MAIZE TECHNICAL MEETING, 45, ANNUAL BLOOD TECHNICAL MEETING, 28, 2000, Pelotas. Proceedings... Pelotas: Embrapa Clima Temperado, 2000. p. 567-573.

GUO, Y. Y. Progress in the researches on migration regularity of *Helicoverpa armigera* and relationships between the pest and its host plants. **Acta Entomologica Sinica**, Beijing, v.40, n.1, p.1-6, 1997.

IMAMT. ***Helicoverpa armigera*: imminent danger to cotton, soybean and corn crops in the state of Mato Grosso**. Technical circular, n.5, 2013, 12 p.

IMEA. **Analysis of the losses and increase in production costs due to the attack by the *Helicoverpa* caterpillar in the 2013/14 harvest in Mato Grosso**. 2013/2014 harvest, Mato Grosso, 2014. 2 p. Available at:

http://www.imea.com.br/upload/pdf/arquivos/Perdas_HELICOVERPA_SAFRA_2013_14_1.pdf. Accessed on: 10/11/2016.

KING, A.B.S. *Heliothis/Helicoverpa* (Lepidoptera: Noctuidae). In:

MATTHEWS, G.A.; TUNSTALL, J.P. (Eds.). **Insect pests of cotton.** Wallingford: CAB International, p.39-106, 1994.

NETO E SILVA, O. A. B.; BOTTON, M.; GARCIA, M. S.; SILVA, A. Da. Effect of growth regulator insecticides on eggs, caterpillars and adults of *Grapholita molesta* (BUSCK) (LEP.: TORTRICIDAE). **Revista Brasileira de Fruticultura**, v.33, n.2, p.420-428, 2001.

PATIL, S.; BASHASAB, F.; VIJAYKUMAR; BASAVANAGOUD; KURUVINASHETTI, M. S.; PATIL, B. V. Genetic relatedness among *Helicoverpa armigera* (Hübner) occurring on different host plants as revealed by random amplified polymorphic DNA markers. **Journal of Asia-Pacific Entomology**, v.9, p.227-233, 2006.

RETNAKARAN, A.; GRANETT, J.; ENNIS, T. Insect growth regulators. **In:** KERKUT, G.A.; GILBERT, L. I. Comprehensive insect physiology biochemistry and pharmacology. New York: Pergamon, 1985. Chap.12, p.529601.

RODRIGUES, R. B.; ARNEMANN, J. A.; FERREIRA, P.; GUEDES, J. V. C.; DALAZEN, G. Efficiency of insecticides in the control of the cotton leafhopper *Alabama argillacea* HUEB., 1818 (LEPIDOPTERA: NOCTUIDAE). **In:** CONGRESSO BRASILEIRO DO ALGODÂO, 7., 2009, Foz do Iguaçu.

Anais... Campina Grande: Embrapa Cotton, 2009. p. 633-638.

RITCHIE, S.; HANWAY, J.J.; THOMPSON, H.E. How a Soybean Plant Develops. Ames, Iowa State University of Science and Technology, Coop. Ext. Serv., 1982. 20p.

RUBIN, L. A. Manejo da Lagarta-do-cartucho *Spodoptera frugiperda* (Lepidoptera, Noctuidae) NA CULTURA DO MILHO: Porto Alegre RS, 2009.
SANTOS, L.M.; REDAELLI, L.R.; DIEFENBACH, L.M.G.; EFROM, C.F.S. Fertility and longevity of *Spodoptera frugiperda* (J. E. Smith).

(Lepidoptera: Noctuidae) in corn genotypes. **Ciência Rural**, Santa Maria, v.34, n.2, p.345-350, mar-abr, 2004.

SILOTO, R.C. Damage and biology of *Spodoptera frugiperda* (J. E. Smith 1797) (Lepidoptera: Noctuidae) in corn genotypes. Piracicaba SP. 2002.

SILVA, D. M.; BUENO, A. F.; MANTOVANI, M. A. M.; STECCA, S. C.; LEITE,

N.; OLIVEIRA, M. C. N. De.; MOSCARDI, F. Biology and feeding preference of *Spodoptera frugiperda* (Lepidoptera: Noctuidae) on different host sources. **In:** VI BRAZILIAN SOYBEAN CONGRESS, 6., 2012, Cuiabà. Proceedings... Cuiabà: Embrapa Cotton, 2012. p. 1-5.

SILVA, M. T. B.; COSTA, E. C.; BOSS, A. Control of *Anticarsia gemmatalis* Huebner (Lepidoptera: Noctuidae) with insect growth regulators. **Ciência Rural**, v.33, n.4, p.601-605, 2003.

SHARMA, P. K.; KUMAR, U.; VYAS, S.; SHARMA, S.; SHRIVASTAVA, S. Monitoring of *Helicoverpa armigera* (Hubner) (Lepidoptera: Noctuidae) through pheromone traps in chickpea (Cicer arietinum) crop and influence of some abiotic factors on insect population. **Journal of Environmental Science, Toxicology and Food Technology**, v.5, n.5, p.44-46, 2012.

SOSA-GÓMEZ, D.R.; SPECHT, A.; PAULA-MORAES, S.V.; LOPES-LIMAC, A.; YANOA, S.A.C.; MICHELI, A.; MORAIS, E.G.F.; GALLOA, P.; PEREIRA, P.R .V.S.; SALVADORIF, J.R.; BOTTONG, M.; ZENKER, M.M.; AZEVEDO-FILHO, W.S. Timeline and geographical distribution of *Helicoverpa armigera* (Hübner) (Lepidoptera, Noctuidae: Heliothinae) in Brazil. **Revista Brasileira de Entomologia**, Curitiba, v.60, p.101-104, 2016.

TOMQUELSKI, G. V.; MARTINS, G. L. M. Efficiency of insecticides on *Spodoptera frugiperda* (J.E. SMITH, 1797) (LEPIDOPTERA: NOCTUIDAE) in corn in the Chapadoes region. **Revista Brasileira de Milho e Sorgo**, v.6, n.1, p.26-39, 2007.

VALICENTE, F. H.; COSTA, E. F. Control of the cartridge caterpillar, *Spodoptera frugiperda* (J.E. Smith), with *Baculovirus spodoptera*, applied via irrigation water. **Anais da Sociedade Entomològica do Brasil**, Jaboticabal, v.24, n.1, p.61-67, 1995.

VIDAL, N. H.; CAMARGO, A. C.; ATAiDE, F. F.; BARROS, R. G.; CZEPAK, C.; SILVEIRA, C. Agronomic efficiency of Flubendiamide 480 SC for the control of *Alabama argillacea* (LEPIDOPTERA: NOCTUIDAE) in the cotton crop. **In:** V Congresso Brasileiro de Algodâo, 2005, Salvador, 2005.

VILELA, E.F.; DELLA LÙCIA, T.M.C. Introduction to Semiochemicals and terminology. In: VILELA, E.F.; DELLA LÙCIA, T.M.C. (Eds.). **Insect pheromones: biology, chemistry and use in pest management.** Ribeirâo Preto: Holos, 2001. p.9-12.

VILELA, M.; MENDES, S.M.; VALICENTE, F.H.; CARVALHO, S.S.S.; SANTOS, A.E.; ARAÙJO, O.G.; BARBOSA, T.A.N.; CARVALHO, E.A.R.; COSTA, V.H.D. **Metodologia para Criação e Manutençao de *Helicoverpa armigera* em Laboratório**. Technical Circular 203, Sete Lagoas, 2014, 7p.

ZALUCKI, M. P. et al. The biology and ecology of *Helicoverpa armigera* (Hübner) and *H. punctigera* Wallengren (Lepidoptera: Noctuidae) **In.** AUSTRALIA: what do we know? *Australian Journal of Zoology*, Melbourne, v.34, n.6, p.779-814, 1986.

ZARBIN, P.H. G.; RODRIGUES, M.A.C.M.; LIMA, E.R. Insect pheromones: technology and challenges for competitive agriculture in Brazil. **Quimica Nova**, Sâo Paulo, v.32, n.3, p.722-731,2009.

WYCKHUYS, K.A.G.; LU,Y; MORALES, H.; VAZQUEZ, L.L.; JESUSA, C.L.; LEGASPI, J.C.; ELIOPOULOS, P.A.; HERNANDEZ, L.M. Current status and potential of conservation biological control for agriculture in the developing world. **Biological Control**, v. 65, p.152-167, 2013.

MIX
Papier aus verantwortungsvollen Quellen
Paper from responsible sources
FSC® C105338

Printed by Books on Demand GmbH, Norderstedt / Germany